Elizia Liauw

Factors affecting nutrition of students and effects of nutrition in educational attainment

GRIN Verlag

Abstract

This paper aims to determine the factors contributing to future generation's educational attainment. It assesses the effectiveness of doing a school-feeding program in increasing school enrollment in the area of Cilincing. This paper attempts to answer the question of: **What are the factors affecting nutrition of student in Sekolah Bambu, Cilincing and how is nutrition affecting educational attainment.**

In the investigation of this question, both qualitative and quantitative data were collected from primary resources. The quantitative data is gained from measuring BMI of 100 random kids, both boys and girls, between the ages of 7 years old to 12 years old. Qualitative data is also obtained through in depth interviews with 10 random families. Both qualitative and quantitative data is gathered from the area around Sekolah Bambu, Cilincing.

This paper arrives at the conclusion that the biggest factor that holds children in Cilincing back from getting sufficient nutrition is essentially because of lack of women empowerment, which could eventually affects educational attainment. Lack of nutrition to an extent does play a role in educational attainment. However, it does not play a big role in determining the number of children that go to school (quantity). It plays a much bigger role in the quality of how students are performing in school, as consuming sufficient nutrients leads to better concentration and cognition. In order to resolve the lack of education for future generations in Cilincing, we must then go to the root of the problem, which is essentially women empowerment.

Word count: 248

Table of Contents

I. <u>Introduction</u>

Many experts believe that sufficient nutrition is one of the keys to increasing one's quality of life. Productivity of a worker on a farm increased at most by 4 percent when his calorie intake increased by 10 percent (Duflo, 27). The benefits of good nutrition might be particularly more important for two sets of people: unborn babies and young children. Children, who receive proper nutrients during their early childhoods, tend to earn more money in the future. Not only that, sufficient nutrition protects children from chronic disease and keeps them healthy. Having a healthy life helps people, particularly the poor, to be even more productive when working and learning.

Cilincing is one of Jakarta's poorest and most densely populated areas. It has one of the highest concentrations of poverty in North Jakarta (Mortensen, Justin.). With this in mind, we wanted to try help the community attain better quality of life by taking the first step of breaking out of the poverty trap. The context of this essay is actually our desire to help people in Cilincing through a school-feeding program in order to increase school enrollment and educational attainment, which in the long run will give them a chance to break out of poverty.

With the idea of this school-feeding program, it led to a bigger unanswered question that this essay will investigate, which is **what are the factors affecting nutrition of student in Sekolah Bambu, Cilincing and how is nutrition affecting educational attainment.**

Investigating on whether nutrition is holding kids back from pursuing education is important because the finding can determine whether or not a school-feeding program plays a role in educational attainment, hence helping them live better lives. And if it is not nutrition that is holding them back, then the investigation could determine what could possibly be done to help people in Cilincing take the first step of breaking out of the poverty trap.

II. <u>Geographical Context</u>

Indonesia has been growing economically in the past decade. However, its growth has been socially and economically uneven. Even though the proportion of people living in poverty fell by half, from 24% in 1999 to 12% in 2012, the gap between the rich and the poor is getting wider ("Muted Music"). According to a forthcoming report by the World Bank, real consumption grew by about 4% a year on average in 2003-10. But for the poorest 40% of households it grew by only 1.3% ("Muted Music").

The presence of this inequality in Indonesia has led to serious health issues, especially in poor households. For instance, the mortality rate for children under 5 is 40 per 1,000 live births—nearly 45% of these child deaths are attributable to various forms of under nutrition (Camila Chaparro). Child mortality rates often reflect poorer households rather than rich households, simply because lack of access to adequate healthcare and nutrition.

III. <u>Location Context</u>

Diagram 1 – Sekolah Bambu, Cilincing ("Jakarta.") (Google Maps: Jakarta, Indonesia [Map data (c) 2015 Google])

As shown in diagram 1, Cilincing is the northeastern most sub-district of Jakarta located at the very far end of Jakarta ("Langgar Tinggi / Rumah si Pitung").

Diagram 2 – Google earth ("Google Maps.") (Google Maps [Map data (c) 2015 Google])

As shown in diagram 2, Sekolah Bambu, which is the place where all data were gathered, is located in the middle of Cilincing. Through this Google earth image of the area, one could conclude that Cilincing is located near the ports and surrounded by industrial areas. Moreover, the residential areas are very dense, especially near the pinpointed location (Sekolah Bambu). This suggests that Cilincing is a relatively poor area with most of the space allocated for industry.

IV. <u>Theoretical Context</u>

Development is more than just economics. It is an expansion of human freedom, capability and opportunity. This could be measure through the Human Development Index (HDI), which measures health, education, and income. All 3 indicators are interrelated to one another and to an extent affect one another's functionality. One way of measuring health is by measuring ones' Body Mass Index (BMI), which is greatly affected by nutrition from food intake. Therefore, when one lacks nutrition and becomes unproductive, it will eventually affect their income, to an extent where one would not be able to work due to malnutrition. Low income families tend to have a hard time in putting their children to school, which leads to situation where the children would grow up working in low paying jobs and have insufficient nutrition due to lack of money. This then creates traps, which many of the poor have a trouble getting out of, called the poverty trap and the health trap.

POVERTY TRAP

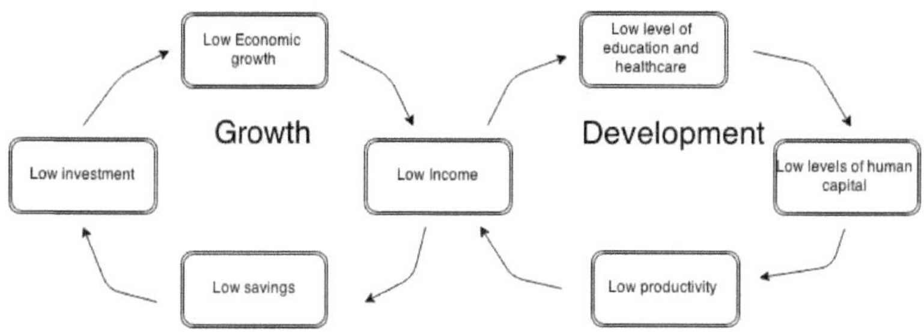

Diagram 3 – The poverty cycle which entraps individuals

As shown in diagram 3, the poverty trap is a never-ending circle with interdependent variables that hinders individual from acquiring better quality of life. Many experts agree that the education of children is one of the strongest keys to break the poverty trap. In many places, Cilincing included, primary education is provided freely by the government. Thus theoretically, economic disability should not prevent kids from attaining education. However, in actuality, parents would prefer their kids to work and help provide for the family.

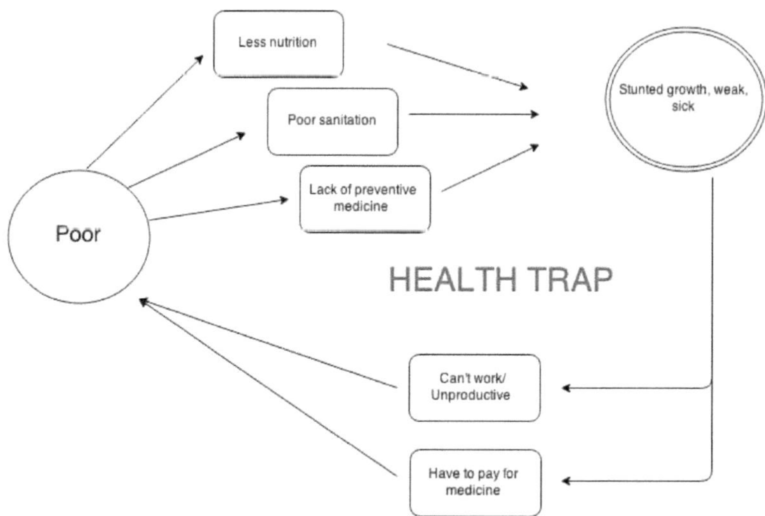

Due to the fact that they have to help provide for the family at a young age and abandon education, the chances of them getting out of the poverty trap is even harder. Moreover, underprivileged families are unable to consume sufficient nutrients because they cannot afford it. Therefore, as shown in diagram 4, lack of nutrients leads to children having stunted growth. This creates a domino effect that will keep them poor, as they become unproductive, causing them to be stuck in a health-trap.

V. <u>Methodology</u>

This research is done by measuring BMI and interviewing families around the area of Sekolah Bambu, Cilincing. Both qualitative and quantitative data were collected from primary resources.

The quantitative data is gained from measuring BMI of 100 random kids, both boys and girls, between the ages of 7 years old to 12 years old. The collected BMI data will be separated into 2 categories, boys aged 7-12 years old and girls aged 7-12 years old, in order to make comparison with standardized BMI index more accurate. This will then be used to identify whether the kids are underweight, normal, or obese. Their BMIs were measured using a technology so it could rule out some calculation errors that might happen. However, the quantitative data might have some slight error because the height and weight measurements were taken manually causing there to be a random error.

Qualitative data is also obtained through interviews with 10 random families, especially the mothers in the area. It is important that these families are randomly selected so that there will be a range of stories, therefore less bias. Interviews are generally better than surveys because they are more personal, as different families have different interesting stories to tell that might help with the investigation. One limitation that occurs is some people that were interviewed might not be totally honest and hide some truth from us for various reasons.

VI. <u>Analysis</u>

1. <u>Food Intake Analysis</u>

All 10 families interviewed have 3 daily diets that they have in common: rice, egg and tempeh. From the interviews, no mothers mentioned any vegetables in any of their diet because all they can afford is rice, egg, and tempeh and only the fortunate enough can afford fish. Earning low and unstable income causes most families in Cilincing to struggle in providing food let alone providing nutritious food for their children. For example, Nursaniah, a mother of 2 young girls, could only afford to feed her two daughters tea for breakfast and egg with rice for lunch (Nursaniah, Interview). They do not eat dinner at all, simply because they cannot afford it.

With most families only having egg and rice in their diet, there is no question that they suffer from nutrient deficiency. Even though they are getting some protein from eggs or tempeh and carbohydrates from rice, they are still not eating enough of them, knowing that most families only eat twice or even once a day. Both tempeh and meat serve as rich sources of complete protein, containing every essential amino acid (Walton, Alice G.). However, tempeh generally contains less protein per serving than meat.

Although some families try to cook and provide food for their children, most mothers decide to just give money to their kids and get their own food. With the freedom that comes with their daily allowance, children tend to buy unhealthy junk food. This daily lack of supervision in 1/3rd or sometimes-even ½ of the growing children's daily intake results in stunting caused by insufficient nutrients in their systems.

By looking at their daily diets and amount of food intake, one could generally conclude that these kids in Cilincing are underweight or even malnourished. Some children are only eating once or twice per day, which could lead them to be malnourished. However, some are getting enough food intakes but the problem is more of the quality of nutrition inside their food intake. To be more definite, stunted growth and malnutrition could be further investigated by looking at BMI measurements.

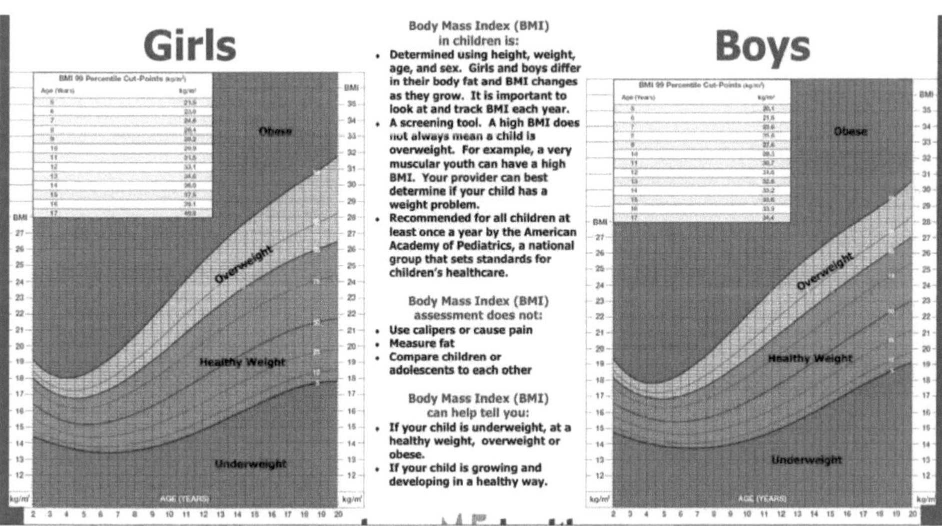

Diagram 5 – Universal BMI Index for children ("BMI Chart For Children.")

This diagram above is an international measurement for BMI standards that shows, for both girls and guys, whether one is underweight, healthy weight, overweight, or obese. The minimum BMI needed, for children with different gender and different age, to be inside the healthy weight zone differs. For boys the minimum BMI required to be considered as healthy weight is 14 kg/m^2 at the age of 7 and it goes up to 15 kg/m2 at the age of 12. For girls it is slightly lower, 13.6 kg/m^2 at the age of 7 and it goes up to 14.6 kg/m^2 at the age of 12.

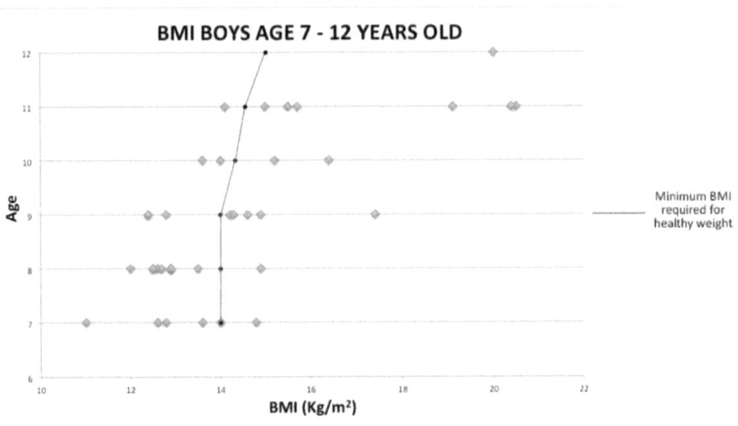

Diagram 6 – BMI Boys in Cilincing (primary resource)

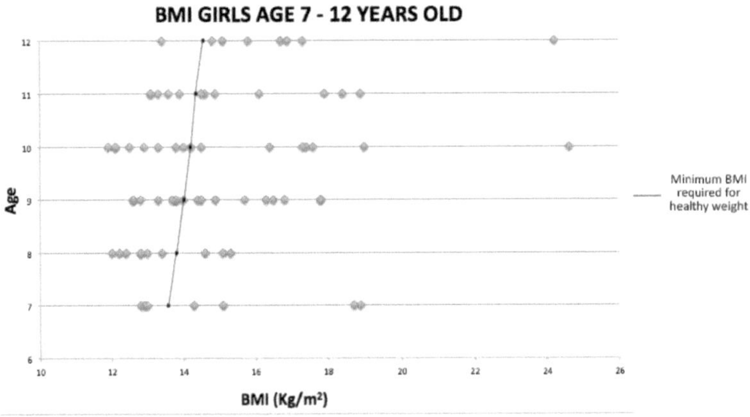

Diagram 7 – BMI Girls in Cilincing (Primary source)

As seen in diagram 6, 50% of the boys measured, as indicated by their respective BMIs, are below the accepted healthy BMI range. In diagram 7, it is shown that there are more girls that are categorized as underweight. This however is due to the fact that there is a greater sample size, as more girls attend the school than boys. It may be true that more girls have BMIs that indicates they are underweight but the percentage is smaller at 37%.

A low BMI, which a large number of the children in Cilincing happen to have, could be traced back to malnutrition. Having a low BMI simply means that the body is not getting enough of the nutrients and energy needed. In children between the ages of 6-12 years, 30 – 45% of their energy is utilized in the brain ("Nutrition and Brain Health | MyVMC."). One of the greatest effects of nutrition on brain functioning is on our cognition. However, lack of sufficient nutrition coming into their body prevents children in Cilincing to power their brain to its maximal capability. To add to that, children in Cilincing are also lacking healthy nutrients in their diet. As mentioned before a large number of children are given money to buy their own food. From the interview gathered, most children in the area will use this money to buy cheap junk food by the streets (Hengky, Interview). People with higher levels of trans fats, such as junk foods, in their blood had poorer performance in thinking, concentrating and in memory tests (Walton, Alice G.).

As shown in diagram 4 (health-trap diagram), stunting caused by insufficient nutrition leads to lack of productivity. Thus, one way to increase working and learning productivity is by getting sufficient nutrients for the body and brain to maximize its functions. Productivity of a worker on a farm increased at most by 4 percent when his calorie intake increased by 10 percent, making nutrition an important factor to increase both working as well as learning productivity (Duflo, 27).

Lack of nutrition to power the brain could also cause other effects that might restrain children from attaining education. From the interviews, some mothers admit that their children sometimes get sick and cannot go to school. Their sicknesses are usually very common and treatable, such as fever, cough, and the flu. However, no specific data of attendance can be gathered because the school does not keep track of them. Other than that, lack of sufficient nutrients in their diet could also lead to attention deficit and lack of focus in classrooms. Deficiencies of certain amino acids, B vitamins and magnesium—all found in vegetables—have been determined to cause significant brain impairment, including short-term memory loss, attention deficit, lack of focus and/or concentration, and mood swings (Helen.). According to Pak Hengky, one of the teachers in Sekolah Bambu, children that do not eat during their break times have a harder time in focusing in class (Hengky, Interview). Thus, theoretically lack of nutrition can be related to attention and focus deficit. However, in actuality this cannot be proven 100% as measuring concentration and focus is very hard, especially with the limited time and resources this research has.

2. Why Aren't They Getting Enough Food?

Indonesia currently has a low level of food insecurity. Food supplies are theoretically adequate to feed the population, however inefficiencies in distribution system across the archipelago restrict access to food products for the nation's poor. As Amartya Sen had shown, most recent famines have been caused not by lack of food availability but by institutional failures that led to poor distribution of the available food (Duflo, 28). According to a research done by Esther Duflo and Abhijit Banerjee, most people, even the very poor, are outside of the nutrition poverty trap zone: they can easily eat as much as they need to be physically productive (Duflo, 38). The problem then becomes less of the quantity of food than its quality, and in particular the shortage of micronutrients in their diets. However for children in Cilincing it might be both. As mentioned before, most families only eat twice or even once a day with diets that lack sufficient nutrients.

The most apparent reason why children in Cilincing are not getting enough nutrition is simply because of the economic state they are in. The results of the families who were interviewed showed an unstable income of roughly Rp 500.000,00 – Rp 2.500.000,00 (approximately $50 - $250) per month. The instability of their income can be caused by their unstable paying jobs as most of the men in the families work as fishermen and scavengers, while the women in the families mostly stay home and take care of their children. For those who work as fishermen, they can only work 6 months every year, due to the fishing season timing. For scavengers, their income varies weekly depending on what they are able to gather that week. This unstable income leads to an unstable diet, where one day they could afford food and other days they might not be able to.

To make matters worse, people in Cilincing tend to have big families even though they have very low income. For example, Ibu Dianah has up to 8 children to feed and only around Rp 1.500.000,00 (approx $150) income per month (Dianah, Interview). Having big families, especially in a poor situation, leads to a quality-quantity tradeoff. When there are more children to feed, each of them will be getting lower quality of food because their parents will devote fewer resources to each child so that they can feed the others. Low quality of food leads to insufficient nutrition consumed, which as mentioned before leads to lack of productivity that traps people in both the poverty and health trap. Children born into large families are less likely to receive proper education, nutrition, and health care, and if poor families are more likely to be large, this creates a mechanism for the intergenerational transmission poverty (Duflo, 108). This then brings to the question why are the families big?

3. <u>Why Do They Have Big Families?</u>

Some reasons why they have big families, especially in Cilincing's situation, are due to limited access to contraception and the common misconception in that area about sterilization reducing the quality of sex. Contraception are provided in the local clinic for about Rp15.000,00 (approx $1.5), but due to the low income and lack of knowledge of how important contraception is, women tend to invest their money in other things (Hengky, Interview). In addition to that, most men prefer their wives to not use any condoms during intercourse because it reduces the pleasure of sex. Not only that, most men in the area also believe that sterilization reduces the pleasure given through intercourse and therefore would rather not have their wife sterilized, even though there is no correlation between sterilization and pleasure given through intercourse. Therefore, from this it could be concluded that in Cilincing men are more dominant in making decisions solely according to their will. This then leads to a bigger problem of gender inequality, which might be one of the root causes of having big families.

As mentioned before, most women in Cilincing stay home as housewives, meaning they do not have any stable income and rely on their husbands. They depend on the money their husbands give them in order to support their family. This dependency gives men more power over their wives, causing them to sometimes discriminate their wives' rights. Most women in the area do not have a stable income mostly due to the lack of education they got as children. Ibu Nursaniah, for example, was forced to leave school at the age of 8 years to work in order to help the family (Nursaniah, Interview). She now works as a scavenger with no stable income. Because of her inability to support herself, she got married at a very young age (16 years old) to a much older man. Women who are much younger than their husbands or much less educated find it harder to stand up to them (Duflo, 116). From this, it can be said that it all comes back to education attainment. When people, especially girls, are educated, they will have a greater chance of having a stable paying job, which will lead them to be less dependent on their husbands. As Lawrence Summers said, "Investment in girls' education may well be the highest-return investment available in the developing world," (Kristof, xx). Not only that, education will also help women acknowledge and fight for their rights. The most effective way to encourage women and girls to stand up for their rights is through education. Therefore, education is the first step to creating a more equal society.

One way of boosting education for the next generation in society is by empowering women rights. By doing this it would create a ripple effect that could eventually fix the problem. When women are valued in a community, more than merely as a housewife, then potentially she could have a voice in decisions made in the household. Through the research that has been done, it can be seen that the barrier of contraception is not necessarily price or lack of access to contraception. It is not as much of an economical problem but more of a social problem. Husbands think that they get to decide whether they should use contraception or not based on their own wills. But with empowering women rights, they have a chance to stand up for themselves. Thus by empowering women, it will actually break down the barrier allowing women to have more authority in the household. Gender equality does not only results in women with more bargaining power in their marriage but also yields a "double dividend," elevating not only women but also their children and communities (Kristof, xx). This is evident, as women would have more authority and bargaining power in the household, resulting in the ability to decide what is best for them. This could lead to having fewer children. Having fewer children will help the family allocate their resources better and allow them to save up some money to invest for their children's education. Nutrition can be a short-term solution to the problem of educational attainment, however women empowerment is the preventative measure that should be taken to tackle the root problem of education attainment, which tackles the problem in the long-term.

VII. <u>Conclusion</u>

From the findings, it can be concluded that having big families and lack of women empowerment affects children's nutrition level immensely.

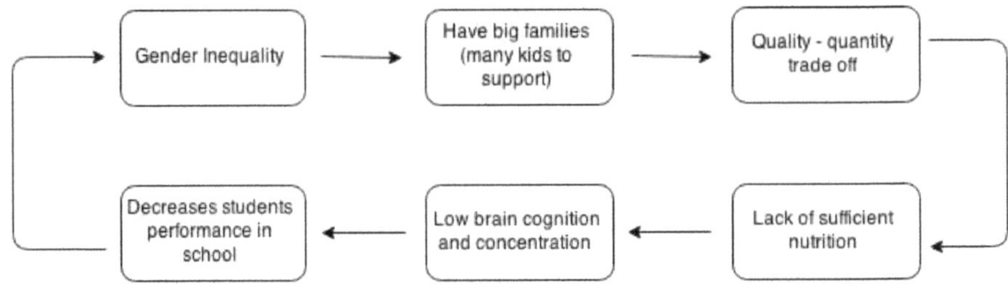

Diagram 8 – Flow Chart

As seen in diagram 8, gender inequality (lack of women empowerment) leads to families having many kids because women do not get to decide what is best for them and is forced to just obey their husbands. Having big families will then lead to a quality – quantity trade off of nutrition. When there are more children to feed, each of them will be getting lower quality of food because their parents will devote fewer resources to each child so that they can feed the others. Lack of sufficient nutrients from low quality of food could potentially hold children back from performing their best at school, as they have low brain cognition and low concentration. However, this correlation between lack of nutrition and low concentration or brain function cannot be tested in Cilincing due to limited resources available. The conclusion can only be drawn through other sources that has done extensive amount of research to make this correlation. If children, especially girls, are held back from their maximum capability, it could lead them to have low paying or unstable jobs in the future. With no stable income, they cannot live independently and have to rely on their husbands, which could lead to a lack of women empowerment and the whole cycle begins again.

Lack of nutrition to an extent does play a role in educational attainment, however it does not play a big role in determining the number of children that go to school (quantity). It plays a much bigger role in the quality of how students are performing in school, as consuming sufficient nutrients leads to better concentration and cognition. All children that were measured do attend school even if they have a low BMI, but those with low BMI might not perform as well as those who are getting more nutrition, which is shown by having higher BMI. However, it is beyond the scope of this investigation and further research is needed to fully examine this statement.

In the case of Cilincing, where there are still a lot of children who do not attend school, it is more important to have bigger quantity of children attending school in order to help them break out of the poverty trap. To resolve the lack of education for future generations in Cilincing, we must then go to the root of the problem, which is essentially women empowerment. Empowering women will give a long term effect in educational attainment. By giving women more authority and power in decision making of the household, families could essentially have fewer children, as women have the power to also decide what is best for her and the family. Fewer children mean that there are more chances for the family to save up and invest on their children's education, which would in the long run increase educational attainment.

Therefore, the biggest factor that holds children in Cilincing back from getting sufficient nutrition is essentially because of lack of women empowerment, which as discussed above has a rippling effect that could eventually affects educational attainment.

<u>**Bibliography**</u>

Banerjee, Abhijit V., and Esther Duflo. *Poor Economics: A Radical Rethinking of the Way to Fight Global Poverty.* New York: PublicAffairs, 2011. Print.

"BMI Chart For Children." *BMI Chart For Children | Search Results | The Works.* The Works, Web. 10 Apr. 2015. <http%3A%2F%2Fthe-works.net%2Ftag%2Fbmi-chart-for-children>.

Camila Chaparro, Lesley Oot, And Kavita Sethuraman. *Indonesia Nutrition Profile.* Washington, DC: National Report on Basic Health Research, Apr. 2010. PDF.

"Cilincing (District)." *Cilincing (District, Indonesia).* Web. 10 Apr. 2015. <http://www.citypopulation.de/php/indonesia-jakarta-admin.php?adm2id=3175060>.

Dianah. "Cilincing Interview". Interview. 29 Mar. 2015.

"Diversity among Economically Less Developed Nations." *Dineshbakashi.* Web. 13 Apr. 2014. <http://dineshbakshi.com/ib-economics/development-economics/169-revision-notes/2133-economic-growth-and-economic-development?tmpl=component>.

"Google Maps." *Google Maps.* Web. 10 May 2015. <https://www.google.com/maps/place/Cilincing,+North+Jakarta+City,+Special+Capital+Region+of+Jakarta,+Republic+of+Indonesia/@-6.1153505,106.9423099,5808m/data=!3m2!1e3!4b1!4m2!3m1!1s0x2e6a20695da8747f:0xc87e8ed840c59afa>.

Hengky. "Cilincing Interview". Personal Interview. 29 Mar. 2015.

Helen. "Could Nutritional Deficiency Be Causing Your Focus Problem?" *Could Nutritional Deficiency Be Causing Your Focus Problem? | The Smart Living Network.* 9 Sept. 2010. Web. 10 Apr. 2015. <https%3A%2F%2Fwww.smartlivingnetwork.com%2Fnutrition%2Fb%2Fis-your-lack-of-focus-actually-nutritional-deficiency%2F>.

"Jakarta." *Google Maps.* Web. 03 Feb. 2014. <https://www.google.com/maps/place/Jakarta/@6.2297465,106.829518,11z/data=!3m1!4b1!4m2!3m1!1s0x2e69f3e945e34b9d:0x5371bf0fdad786a2>.

Kristof, Nicholas D., and Sheryl WuDunn. *Half the Sky: Turning Oppression into Opportunity for Women Worldwide.* New York: Alfred A. Knopf, 2009. Print.

"Langgar Tinggi / Rumah si Pitung". *Dinas Pariwisata Dan kebudayaan Provinsi DKI Jakarta* (in Indonesian).
<https://www.jakarta.go.id. Retrieved April 13, 2010.>.

Mortensen, Justin. "Raising Awareness of Urban Poverty on World Cities Day." *'Save the Children Voices from the Field'* 31 Oct. 2014. Web. 10 Apr. 2015.
<http://savethechildren.typepad.com/blog/2014/10/save-the-children-supports-the-first-world-cities-day.html>.

"Muted Music." *The Economist*. The Economist Newspaper, 03 May 2014. Web. 10 Apr. 2015.
<http://www.economist.com/news/finance-and-economics/21601550-poor-are-benefiting-relatively-little-indonesias-growth-muted-music>.

"Nutrition and Brain Health I MyVMC." *MyVMC*. 13 Jan. 2010. Web. 10 Apr. 2015.
<http://www.myvmc.com/lifestyles/nutrition-and-brain-health/>.

Nursaniah. "Cilincing Interview". Interview. 29 Mar. 2015.

Walton, Alice G. "The Connection Between Good Nutrition and Good Cognition." *The Atlantic*. Atlantic Media Company, 13 Jan. 2012. Web. 10 Apr. 2015.
<http://www.theatlantic.com/health/archive/2012/01/the-connection-between-good-nutrition-and-good-cognition/251227/>.